INSTITUTE OF PETROLEUM

Guidelines on the Use of Oil Spill Dispersants

INSTITUTE OF PETROLEUM

Guidelines on the Use of Oil Spill Dispersants

Second Edition

August 1986

Published by
John Wiley & Sons Limited
on behalf of
THE INSTITUTE OF PETROLEUM, LONDON

Copyright © 1987 by the Institute of Petroleum, London

All rights reserved.

No part of this book may be reproduced by any means, or transmitted, or translated into a machine language without the written permission of the publisher.

British Library Cataloguing in Publication Data:

Guidelines on the use of oil spill
  dispersants.—2nd ed.—(IP code of
  practice for metal working fluids)
  1. Oil pollution of the sea 2. Dispersing
  agents
  I. Institute of Petroleum II. Code of
  practice for the use of oil slick
  dispersants III. Series
  628.1′6833′09162   GC1085

ISBN 0 471 91405 3

Acknowledgement: Figs C1–C10, Crown Copyright. Reproduced by permission of the Director, Warren Spring Laboratory.

Printed in Great Britain by Galliard (Printers) Ltd, Great Yarmouth

CONTENTS

Foreword		vii
Oil Dispersants Working Group Membership		ix
Section	**1 Introduction**	1
	2 Application of Dispersants	4
	2.1 Sea	4
	2.2 Shorelines	5
	2.3 Man Made Structures	5
	2.4 General Comment	6
	3 Limitations on the Use of Dispersants	7
	3.1 Physical Limitations	7
	3.2 UK Legislation	8
	3.3 Environmental Effects	9
	4 Safety, Handling and Storage	10
	4.1 Safety In Use	10
	4.2 Safe Handling	10
	4.3 Storage	10
	4.4 Disposal	11
	References	12
Appendix A	**Specification for Oil Spill Dispersants—Performance**	13
B	**Specification for Oil Spill Dispersants—Toxicity**	21
C	**Application Equipment for Dispersant Concentrates**	26
D	**Authorities Approving the Use of Dispersants in UK Waters**	32
E	**Bibliography**	33
	Suggestions for Further Reading	33

FOREWORD

The first edition of this booklet was published in 1979 under the title 'Code of Practice For The Use of Oil Slick Dispersants'. Since that time there has been a great deal of research and development work on new methods to reduce the effects of oil pollution of the sea. A substantial proportion of this work has been directed towards oil spill dispersants and their methods of application, particularly aerial spraying. This has resulted in the formulation of high efficiency concentrated dispersants and compatible aerial spraying systems thereby significantly improving the response time to any given oil spill.

This revised edition of the booklet renamed 'Guidelines' takes account of these developments and it is based largely on experience and legislation relevant to the UK. The general principles will apply to most countries of the world but it must be borne in mind that different regulations covering the application, toxicity and performance requirements of dispersants will often apply outside the UK. These Guidelines must be regarded as complementary to such regulations and must not be interpreted as abrogating them.

Whilst the adoption of these Guidelines will assist in combating oil pollution of the sea with dispersants in an efficient manner, the IP cannot accept any responsibility for injury to persons or damage to or loss of property however arising out of, or in conjunction with, the application of or reliance upon these Guidelines.

July 1986

R. J. FARN
Chairman
IP Dispersants Working Group

INSTITUTE OF PETROLEUM, MARINE ENVIRONMENT COMMITTEE
MEMBERSHIP OF DISPERSANTS WORKING GROUP, JULY 1986

Mr R. J. Farn	BP Detergents International (Chairman)
Mr J. L. Belk	Dasic International Ltd
Mrs F. L. Franklin	Ministry of Agriculture, Fisheries and Food (Part Time)
Dr D. E. Kent	Petrofina (UK) Ltd
Mr A. Lewis	BP Research Centre
Mr R. Lloyd	Ministry of Agriculture, Fisheries and Food
Mr B. W. J. Lynch	Department of Transport, Marine Directorate (MPCU)
Mr J. A. Nichols	International Tanker Owners Pollution Federation Ltd
Mr P. R. Morris	Warren Spring Laboratory
Mr J. K. Rudd	Amoco Europe Incorporated
Mr D. I. Stonor	Shell International Petroleum Company Ltd
Mr R. J. Sutton	Essochem Performance Chemicals Ltd

Mr R. J. Donally, Nature Conservancy Council gave advice on Section 3.3

1

INTRODUCTION

These guidelines are intended to assist those who use dispersants to treat oil spills in the marine environment.

The key to the successful use of dispersants is speed of response since many oils rapidly become resistant to treatment due to weathering. A comprehensive contingency plan is therefore required which clearly states the case for using dispersants in different locations so that equipment and materials can be sited and stored accordingly.

The use of dispersants is often the only practical means of removing spilt oil from the surface of the sea thereby protecting resources in its path such as amenity beaches, shellfisheries and wildlife. Containment and collection of oil, although in principle a more desirable approach, is often impractical, particularly in open waters. However, dispersants also have their limitations in that they do not work very well in calm conditions or against viscous oils and weathered water-in-oil emulsions. When a contingency plan calls for both dispersant and mechanical collection methods to be considered, a decision tree, such as that depicted in Fig. 1, may be useful.

All dispersants work by changing the interfacial properties of oil and water so that mild agitation will break-up the oil layer more easily into very small droplets which will be dispersed in the upper layers of the sea thus lessening the physical impact of the oil and making it more available for biodegradation. The process is shown schematically in Fig. 2.[1]

The potential effects of chemically dispersed oil on marine life depend on where the oil has been spilt and the type of organism present. The probability is that the use of low toxicity dispersants offshore will not cause significantly more damage to marine life than if the oil were not dispersed. However, in shallow inshore areas and on the foreshore where the scope for rapid dilution of dispersed oil is less and where more sensitive resources (e.g. shellfisheries, fish nursery areas) may be at risk, the use of dispersants may result in more damage than if the oil were left alone.

Government regulatory agencies in many countries require dispersants to meet certain standards of efficiency and toxicity before being approved for use. Three types of low toxicity products are now commercially available.

Type 1 which are hydrocarbon solvent based and should be applied undiluted to oil spills.

Type 2 concentrates which are normally diluted with sea water at a ratio of 1:10 prior to application.

Type 3 high efficiency concentrates which are applied undiluted to oil spills.

The Type 3 concentrates have largely superseded the hydrocarbon based products and water dilutable concentrates (Types 1 and 2), particularly for dispersion at sea because they are more efficient when applied undiluted and their concentrated nature increases the time for spraying before replenishment of stocks. These characteristics have also led to the development of application techniques using aircraft. Apart from the advantages of rapid response and high treatment rate, the use of aircraft has proved more effective than other spraying methods, especially in the case of large spills.

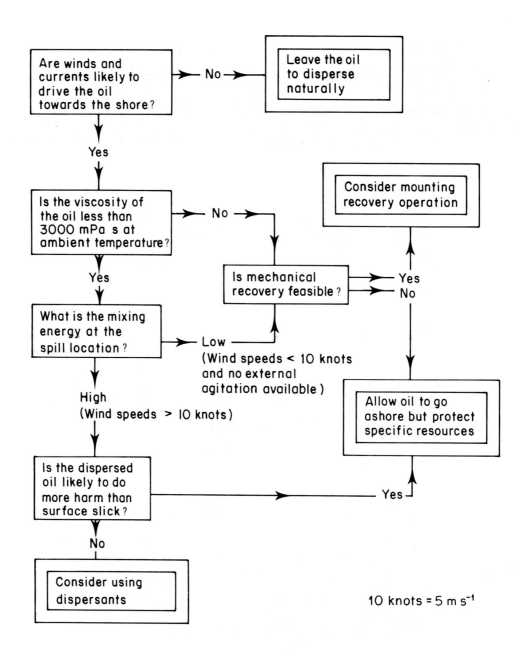

Figure 1. Dispersant usage decision tree.

INTRODUCTION

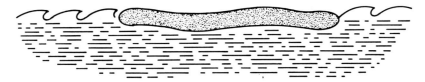

(A) Oil spill

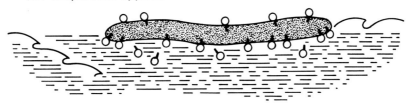

(B) Dispersant applied

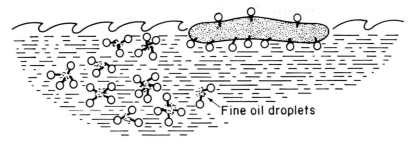

(C) Agitation forms oil droplets

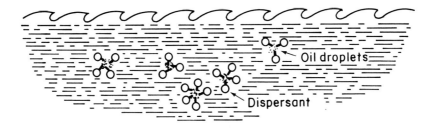

(D) Dispersant prevents coalescence of droplets

Figure 2. Dispersion of typical oil spill.

2

APPLICATION OF DISPERSANTS

2.1 SEA

In making the decision to use dispersants (Fig. 1) the following points must be considered:

(a) The spilt oil must be of a type which is amenable to dispersant treatment (see Section 3.1—Limitations to Use).
(b) The dispersant must be applied to the surface oil at the correct dose rate and in the correct manner.
(c) Sufficient mixing energy must be available to remove dispersant/oil mixture from the surface and disperse it into the body of the sea.
(d) Oil which is initially amenable to dispersant treatment can become resistant to treatment over a period of time, due to weathering processes (see Section 3.1—Limitations to Use). Therefore, speed of response is crucial in providing effective dispersant treatment.

The limitations imposed on dispersant effectiveness by oil type and weathering processes will be discussed in more detail in another section (Section 3.1). The other factors enumerated above are now discussed in general terms in order to emphasise some of the problems associated with the operational use of dispersants to treat oil spills at sea.

2.1.1 Distribution of Oil

When oil is spilt at sea it will spread over the surface. However, because of the action of both wind and tide, the layer of oil within the affected area will not be uniform but consist of fragmented patches of thick oil surrounded by large areas of sheen. In a sea trial[2] the range of oil thicknesses present in the total area covered by an oil spill were found to vary by at least 5 orders of magnitude, i.e. from 1×10^{-8} metres up to 1×10^{-3} metres (1 millimetre). Because of the range of oil thicknesses which may be encountered within an oil slick it is not possible to specify precise application rates for the overall treatment of the affected area. However, treatment of the thicker patches of oil within the 1×10^{-3} metres thickness range will provide a higher clearance rate, in terms of the quantity of oil treated per unit time, than treatment of those areas in which the oil is present in film thicknesses of about 10^{-8} metres. Therefore in an oil spill which may cover several square kilometres it is important that these thicker patches, containing the largest quantities of the spilt oil, should be located and treated effectively by the spraying units in order to maximise the clean-up operation at sea.

2.1.2 Application, Dose Rate and Mixing Energy

The floating oil is treated by means of spraying systems which apply the dispersant in the form of an even spray of droplets to the oil layer. The spraying systems (see Appendix C) are designed to produce droplets of a large enough diameter to resist the effect of wind drift but not so large that they will pass completely through the oil layer and be lost to the underlying water (see Section 3.1.2).

Until the early 1980's, the technique for dealing with oil at sea was based largely on spraying from surface vessels. These systems used either hydrocarbon solvent type or concentrate dispersants diluted 1:10 with sea water prior to application. The resulting oil/dispersant mixture was mixed into the underlying sea water by means of agitation boards towed behind the spraying vessel.

Subsequent research[3] showed that concentrate dispersants were more effective when applied undiluted to floating oil at a dispersant to oil ratio (DOR) of 1:20 and it was considered that the natural action of the sea would provide sufficient mixing energy to disperse the treated oil. The increased effectiveness of the concentrate dispersants, their low mixing energy requirement when applied undiluted and the necessity to treat the spilt oil before the weathering processes made it resistant to treatment led to the development of aircraft spraying systems using undiluted concentrate dispersants. The use of fixed wing aircraft

APPLICATION OF DISPERSANTS

provided a rapid response system to locate and rapidly treat those thicker patches of oil, within the affected area.

Since 1983, the main response by the UK Government to an oil slick threatening the coastline has been to spray the oil using fixed wing aircraft carrying Type 3 concentrate dispersants and fitted with suitably designed spraying equipment. Details of these aircraft systems are included in Appendix C. During an oil spill incident, the specific task of spraying aircraft is to locate and treat the thicker patches of oil. Because aerial treatment of oil requires accurate targeting of the dispersant with the aircraft flying at a low level above the sea surface, it is considered that spraying aircraft should work in pairs with one remaining at a high level to provide guidance to the aircraft engaged in spraying.

Concurrent with the use of fixed wing aircraft spraying systems, helicopter spray bucket attachments have been developed for aerial treatment of floating oil at sea. Additionally, in the last two years, ship spray systems for the application of undiluted Type 3 concentrate dispersants have been developed. Full details of all these spraying systems are also included in Appendix C.

Whichever system is used, the operational objectives in mounting an at-sea response are:

(a) Early treatment of the oil with undiluted high efficiency concentrate dispersant (Type 3).
(b) Location of the thicker patches of oil within the oil slick.
(c) Effective treatment of these thicker patches at the optimum DOR.

In using the ship mounted systems location of the thick patches of oil may be difficult because of the low visual range afforded by surface vessels. Therefore, the use of a spotter aircraft or helicopter with trained and experienced observers will be required to control surface vessel spraying operations.

2.2 SHORELINES

2.2.1 Sandy Beaches

Dispersants, preferably Type 3, may be used to disperse stranded oil from sandy beaches provided that the oil layer is less than 6 mm thick. In instances where the beach is heavily contaminated, dispersants may be used for a final clean up after the gross pollution has been removed. Dispersants are applied to the stranded oil in advance of the flooding tide allowing at least 0.5–3 hours contact time before the water reaches the treated oil. For the oil to disperse, it must be in contact with the sea water. This may require treated oil which is above the existing high tide line to be pushed down to a lower beach level.

Good dispersion also requires sufficient mixing energy which is normally provided by the surf action.

2.2.2 Shingle Beaches

Oiled shingle can be treated in a similar manner to sandy beaches. Early cautionary statements about treated oil penetrating down into shingle beaches are no longer considered valid provided Type 3 dispersants are applied on the incoming tide. Under these conditions maximum advantage may be taken of the higher wave energy levels associated with such beaches and the oil cannot penetrate below the water level in the beach.

2.2.3 Rocky Shores

The wave energy associated with this type of coastline will provide a high natural cleaning rate. Therefore rocky areas of coastline polluted with oil should be left, if possible, to allow the oil to degrade naturally. However, if cleaning is considered necessary, dispersants may be used on the thin layers of oil deposited on the rock faces. In areas such as rock pools, where the oil may be deposited in very thick layers, it should be physically removed.

Dispersant treatment of the thin layers should be carried out in advance of the incoming tide. Sufficient energy to disperse the treated oil may be supplied by the flooding tide or alternatively by high pressure hosing. Particular care should be taken to ensure that dispersant does not accumulate in rock pools or remain there for long periods of time.

2.2.4 Mud Flats

These tend to be located in areas of low wave energy such as bays, inlets and estuaries. Although the natural degradation of stranded oil will be a slow process in this low energy environment, dispersant treatment should be avoided. This is because there will be insufficient energy to disperse the treated oil and flush it out of these areas which may contain fish spawning grounds or house shellfish beds.

2.2.5 Salt Marshes

Almost any clean-up approach is likely to increase damage rather than decrease it and dispersants should not be used in these areas.

2.3 MAN MADE STRUCTURES

Oiled sea walls and man made structures may be cleaned by dispersant treatment. However, on large vertical areas normal dispersant treatment has the disadvantage of a fast 'run off' of the liquid resulting in a very short contact time. This problem may be overcome by the use of a dispersant gel mixture. This material, in a heavy gel form, may be applied to vertical surfaces and left for many hours in contact with the oil, slowly leaching the dispersant into the oil. The equipment for applying dispersant gel (see Appendix C) consists of twin pressurised containers supplying the materials via flexible tubing to a spray lance. When the sea is calm a high pressure hose will assist the removal of the oil. In some instances it may be more convenient to remove the oil by high pressure cold water/hot water/steam washing but it should be remembered that oil removed by these means will not disperse and will have to be recovered.

2.4 GENERAL COMMENT

In clean-up operations where dispersants are to be used close inshore, both the Local Fisheries Officer and the Nature Conservancy Council should be consulted (see Section 3.3).

3

LIMITATIONS ON THE USE OF DISPERSANTS

The use of oil spill dispersants is carefully controlled by regulations (Section 3.2). Separately to the regulatory control, there are situations where the physical condition of the oil and the prevailing environmental conditions make it inappropriate to use dispersants, since they may be ineffective in the situation. These so called limitations are now discussed below.

3.1 PHYSICAL LIMITATIONS

3.1.1 Oil and Emulsion Viscosity

The behaviour of oil when spilt is to spread rapidly over the sea to form a thin layer from which the volatile components will evaporate. At the same time an emulsion of seawater in the oil layer is formed. These phenomena lead to a rapid increase in the apparent viscosity of the oil and a change in its appearance from black through to dark brown and to orange.

It has been found that the efficiency of a dispersant applied at sea to such a weathered oil falls away as viscosity rises above about 3000 mPa s (see Fig. 3). Higher viscosities are amenable to treatment on beaches. One reason for the loss of efficiency with increasing viscosity is because the rate of diffusion of the active ingredients of a dispersant is viscosity dependent. Unless they can be absorbed into the oil film quickly they are lost into the sea by wave action. Consequently, before dispersant spraying is undertaken, every effort needs to be made to verify, by sampling or by a knowledge of the characteristics of the oil spill, that the viscosity of the oil at the time of treatment is still below about 3000 mPa s at sea temperatures.

3.1.2 Dispersant Droplet Size

There is an optimum droplet size for undiluted dispersants (Type 3) of between 500–700 μm (mean volume diameters) for applying the dispersant to the oil. The use of equipment producing droplets outside this range may be wasteful of dispersant. The droplets less than 500 μm size may be blown away by the wind and lost, while large droplets much above the 700 μm size can pass through the oil film and be lost into the sea. In order to minimise possible wastage of dispersant, it is advisable to release the dispersant as close to the surface of the oil spill as practical. In the case of aircraft applying undiluted dispersant, they should fly as low as possible above the sea to be effective. Application of dispersant at too high an altitude can also result in some solvent, added to the dispersant to improve its compatibility with oil, being lost by evaporation.

3.1.3 Sea Conditions

The use of neat dispersant applied from an aircraft or from a boat without external mixing, e.g. breaker boards, will only be effective when there is sufficient wave energy in the sea. The formation of a dispersion relies on the current shear in the sea to create droplets sufficiently small to become widely distributed without resurfacing and coalescing.

The minimum energy needed will vary with the apparent viscosity and nature of the oil spilt. In order to avoid wastage of dispersant it is desirable to be certain that the surface of the sea is agitated preferably with breaking waves present. The use of dispersants when the sea is calm and the surface winds are less than about 10 knots (5 m s^{-1}) is not recommended. Dispersants applied in calm conditions are likely to diffuse into the sea and be lost without any dispersion of oil taking place.

3.1.4 Salinity

The formation of small, stable droplets of oil which migrate with the currents and do not resurface and

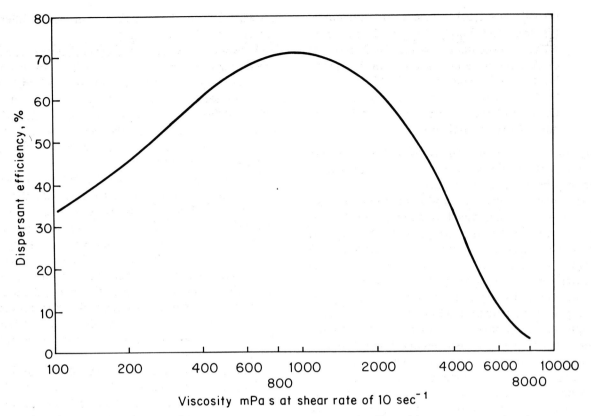

Figure 3. Typical dispersant efficiency (10°C) curve for crude oil emulsion vs apparent viscosity.

coalesce relies on the presence of an electrolyte (salt) in the water. Consequently, most dispersants are less effective in fresh water or in low-salinity sea water. The use of fresh water from fire hydrants is not an effective way to wash away oil on amenity beaches after treatment with dispersant.

3.1.5 Low Temperature Effect

Low ambient air temperature reduces the rate of diffusion of the dispersant into the oil film and at the same time increases the viscosity of the oil. Both these reduce the effectiveness of the dispersant, but the greater effect by far is that of the oil viscosity which seems to outweigh small differences in rates of diffusion at different temperatures.

Low ambient air temperature also increases the viscosity of the dispersant and difficulty may be encountered in spraying the dispersant at the required treatment level and with the correct droplet size distribution. Where the ambient air temperature is persistently less than 0°C, some dispersants (see manufacturer's instructions) should be kept warm before application.

3.1.6 Relative Density of the Oil

The ability of an oil droplet to remain dispersed in the sea is governed by a number of factors, including the relative density of the oil. It is known that, subject to the over-riding viscosity limitation, it is easier to disperse the more dense products; this is illustrated by the curve shown in Fig. 3, where it will be seen that the efficiency falls away from its maximum at lower viscosities when density is also reduced.

3.2 UK LEGISLATION

From 1974 to 1985, the use of products manufactured for the treatment of oil spills was controlled under the Dumping at Sea Act 1974 (DASA) which required both the product and the user to be licensed.

Under the provisions of the Food and Environment Protection Act 1985, Part II (which replaced DASA on 1 January 1986), and paragraph 21 of the Deposits in the Sea (Exemptions) Order 1985, control is exercised by approval of the products. The user does not require a licence provided that the following conditions are fulfilled.

(a) the product used must be one which is currently approved by the licensing authority;

(b) it must be used in accordance with any conditions to which the approval was subject; and

(c) must not be used in an area of the sea of a depth of less than 20 metres or within 1 mile of any such area, save with the approval of the licensing authority (in practice, this approval is obtained in England and Wales through the local Ministry of Agriculture, Fisheries and Food's District Inspector of Fisheries; in Scotland through the Department of Agriculture and Fisheries for Scotland's local Fisheries Officer; and in Northern Ireland, through the Environmental Protection Division of the Department of the Environment, Northern Ireland).

Any user who fails to comply with these statutory requirements would be in breach of the regulations and may be liable to prosecution.

Approvals of individual products used in the treatment of oil spills, are issued to manufacturers on the basis of data from tests to assess both efficiency and safety as well as the toxicity of the product to marine organisms. The product must first be submitted to the Warren Spring Laboratory (WSL) for an assessment of efficiency and safety against the specification given in Appendix A. If these criteria are satisfied, a sub-sample of the product is passed by WSL to the Ministry of Agriculture Fisheries and Food, Fisheries Laboratory at Burnham-on-Crouch where the toxicity to marine organisms is assessed by tests described in Appendix B to determine its acceptability for use at sea, on sandy/pebble amenity beaches and on rocky shores, as appropriate. A full list of the products currently approved can be obtained from any of the addresses in Appendix D or from Warren Spring Laboratory.

3.3 ENVIRONMENTAL EFFECTS

The major environmental effects of an oil spill are derived from its physical properties, for example the oiling of sea birds or the smothering of organisms on rocky shores, or chemical toxicity caused by natural dispersion of the oil into the sub-surface water, where the droplets and soluble fractions can harm fish and shellfish. Early dispersant formulations were very toxic to marine organisms and their use added significantly to the toxicity of dispersed oil. Modern dispersants, however, are much less toxic and, although they have a measurable toxicity, when properly used they should not add significantly to the toxicity of the oil. However, although oil spill dispersants are relatively safe, there remain the problems associated with the transfer of oil to a different environmental compartment.

A decision may have to be made, therefore, as to whether spilt oil should be left as it is, for example on the sea surface (where it can harm sea birds) on a rocky or sandy shore or salt marsh (where it can affect wildlife and amenity) or whether it should be chemically dispersed into the sea where it will be diluted and degraded, but where in shallow waters it could harm fish and shellfish.

This decision depends on the balance between the environmental damage associated with the two options. If sea birds or sensitive coastal resources are threatened by oil pollution, the use of dispersants where the water is deep, or the water exchange is great, is likely to be the most effective way to protect them, and any harm to other organisms in such sea areas will be minimal. On the other hand, the use of dispersants to remove oil from a rocky shore may add to the damage already done to the flora and fauna there, and the best course may be to leave the oil to weather naturally.

In shallow sea areas and on the shore-line, there may be other factors to take into account. For example, where oil is stranded on sandy shores or salt marshes, dispersants should not be used if there is a possibility that the dispersed oil will penetrate below the surface, as buried oil has a very slow degradation rate and its toxic effects may be persistent. However, dispersant usage is acceptable if it enables the dispersed oil to mix and dilute in the waters of an incoming tide and if there are no significant resources at risk immediately above or below the low tide mark.

The relative importance of the various inshore resources—amenity, fishery, nature conservation etc—will change seasonally and where possible this should be allowed for and included in detailed contingency plans drawn up in advance of a spill occurring. Nevertheless, because each oil spill situation is unique, no contingency plan can cover every permutation and combination of potential effects, and it is imperative that expert advice be sought from the local representative of the appropriate Government Department (see section 3.2) and the Regional Officer of the Nature Conservancy Council as soon as possible, so that the environmnental consequences of using the various response options available can be assessed.

4

SAFETY, HANDLING & STORAGE

Because individual dispersants vary in their physical and chemical characteristics, the information given in this section can be of a general nature only. In all circumstances those intending to use, store or handle dispersants should consult the relevant supplier for specific guidance on appropriate precautions and practices.

4.1 SAFETY IN USE

Dispersants contain emulsifiers and solvents with a strong degreasing action and certain elementary precautions must be taken to protect the skin, eyes and respiratory tract from spray.

Those involved in spraying operations should protect themselves by wearing:

- Full cover plastic overalls
- PVC gloves
- Closely fitting face visor
- Chemical resistant safety footwear

The visor would normally be worn as an attachment to a safety helmet. Chemical safety goggles when worn with a closely fitting mouth and nose mask may be considered as an alternative to a visor.

4.2 SAFE HANDLING

Standard precautions for the handling of industrial chemicals should be observed, while the same protective gear as mentioned in 4.1 is applicable. Handling should be undertaken only in well ventilated areas away from heat, sparks and open flames. Inhalation of product vapours should be avoided.

Any accidental skin or eye contact by dispersant should be treated by prolonged fresh water flushing, followed by medical attention if irritation persists.

4.2.1 Fire Hazard

Many dispersants are combustible and normal precautions should be taken. Fires should be extinguished using carbon dioxide, chemical powder, foam, sand or earth. Additionally, larger fires may be fought using water fog.

4.2.2 Dispersant Spillage

Spilt dispersant creates a slippery and therefore unsafe work area and spills should be minimised by the promotion of good housekeeping practices. Dispersants also have a deleterious effect on many paint coatings.

In case of spillage, dispersant should be mopped up with suitable sorbent and the area flushed with copious quantities of water. For larger spillages the dispersant should be prevented from spreading or entering drains by using any available barrier (e.g. sand or earth), before collecting into a tank, absorbing residual quantities and finally flushing with water. It is important not to allow substantial quantities of dispersant to drain from the spill site, in order to avoid polluting surface waters or sewage treatment plant.

4.3 STORAGE

Dispersants are frequently stored in standard sealed steel or plastic drums, as supplied by the manufacturers. They are also stored in vented bulk tanks, usually of mild steel, either ashore or on board vessels.

Dispersants should be maintained at an even temperature, preferably in the range $-10°C$ to $+30°C$. If stored in this manner in sealed containers, the product can be expected to retain its original properties, on which approval was based, for a period of at least two years (see Appendix A).

SAFETY, HANDLING & STORAGE

If dispersants are stored under conditions where they are accessible to air and moisture, such as in vented bulk tanks, there is some evidence to suggest that in some cases deterioration of the product may be accelerated and its useful shelf life reduced.

To satisfy themselves that any contingency stocks of dispersant are of an adequate quality, stockholders and operators are recommended to arrange periodic testing, say annually, of random samples in accordance with the efficiency test in Appendix A. Where storage tanks are involved it is suggested that samples be taken from top and bottom and tested separately.

Different dispersants of the same qualifying 'Type' (see Appendix A) should not be mixed without prior consideration of their compatibility. Dispersants of different qualifying 'Types' should never be mixed.

4.4 DISPOSAL

Should a stock of dispersant no longer be fit for its intended purpose, it must be disposed of in an environmentally acceptable way. There are however several potential options, some of which provide beneficial use. The options include:

- Reprocessing by the original supplier or a specialist
- Use for deck, tank or heavy duty industrial cleaning (having regard for relevant standards of the discarded washwater)
- Blend into fuel oil for heat recovery (detailed technical advice is essential)
- Incinerate or bury at an approved site (in strict compliance with all applicable legislation).

REFERENCES

1. Canevari, G. P. 'General Dispersant Theory' Proceedings of Joint Conference on Prevention and Control of Oil Spills, API/FWPCA, New York City, NY, Dec. 1969.
2. Cormack, D. The Use of Aircraft for Dispersant Treatment of Oil Slicks at Sea—Report of a Joint UK Government/Esso Petroleum Company Limited, Investigation. Marine Pollution Control Unit, Department of Transport, London (1983).
3. Martinelli, F. N. and Cormack, D. Investigation of the Effects of Oil Viscosity and Water-in-Oil Emulsion Formation on Dispersant Efficiency. Stevenage—Warren Spring Laboratory 1979, Report LR 313 (OP).

APPENDIX A

SPECIFICATION FOR OIL SPILL DISPERSANTS—PERFORMANCE

1 SCOPE

This Specification relates to oil spill dispersants for application at sea or on beaches and to the procedures adopted in the UK before any dispersant can be so used. Full details may be found in Institute of Petroleum publication IP 83-009.

2 PROCEDURES

(a) *Qualification*

A dispersant which has been found to conform to the requirements of the Specification when subjected to tests carried out by the Qualification Authority, Warren Spring Laboratory (WSL), will be granted a Certificate of Qualification, if the information required in Clause 5 below has been provided.

A Certificate of Qualification will normally be valid for a period of five years from the date on which it is granted, during which time the dispersant may be referred to as a 'Qualified Product'. It will be necessary for a further sample of the product to be submitted to the Qualification Authority six months before the end of the period of validity if it is desired to extend the period of Qualification.

(b) *Approval for Use*

No dispersant may be used on UK beaches or in 'UK waters', as defined by the Food and Environmental Protection Act 1985, unless its use has been approved by the Minister of Agriculture, Fisheries and Food (MAFF), the Secretary of State for Wales, the Secretary of State for Scotland or the Department of the Environment for Northern Ireland (DoE(NI)). The Ministry of Agriculture, Fisheries and Food acts on behalf of the departments concerned in Wales, Scotland and Northern Ireland. Approval will only be issued in respect of a product which (i) is the subject of a valid Certificate of Qualification and (ii) has been found to conform to the MAFF requirements concerning toxicity to marine organisms, in tests conducted by MAFF on a sample of the product supplied by WSL.

Note: The authority of the UK authorities to approve oil spill dispersants relates to UK waters and UK ships and aircraft only. Dispersants should only be considered suitable for use outside the UK when the agreement of the appropriate National Authority has been obtained even if carried by a UK ship or aircraft and a UK approval is held. A list of products approved by the UK authority can be obtained from either WSL or MAFF.

(c) *Charges and Samples for Qualification and Approval*

The Qualification and Approval Authorities impose separate charges for the work involved in their respective activities. These charges must be paid by the person or organisation submitting a product for Qualification or Approval. That person or organisation must also provide, free of charge and carriage paid, such samples of the product as are required by either Authority.

GUIDELINES ON THE USE OF OIL SPILL DISPERSANTS

(d) *Correspondence*
Correspondence concerning Qualification should be sent to:

The Director
Warren Spring Laboratory
Gunnels Wood Road
Stevenage
Herts SG1 2BX

(Tel: Stevenage (0438) 313388; Telex No. 82250)

and for Approval to:

The Ministry of Agriculture, Fisheries and Food
Fisheries Division 1C
Marine Pollution Branch
Great Westminster House
Horseferry Road
London SW1P 2AE

(Tel: 01-216 6311; Telex No. 21271)

Toxicity testing is carried out by the MAFF Fisheries Laboratory, and any queries on technical aspects of approval, including toxicity testing, should be addressed to:

The Ministry of Agriculture, Fisheries and Food
Fisheries Laboratory
Remembrance Avenue
Burnham-on-Crouch
Essex CM0 8HA

(Tel: Maldon (0621) 782658; Telex No. 995543)

3 TYPES OF DISPERSANT

This Specification relates to three types of oil spill dispersant. These are.

Type 1 Conventional, Hydrocarbon-base—for use primarily undiluted on beaches, but may also be used undiluted at sea from WSL spray sets using breaker boards or other suitable means of application and agitation.

Type 2 Water-Dilutable Concentrate—for use at sea after dilution 1:10 with sea water and sprayed from WSL spray sets using breaker boards or other suitable means of application and agitation.

Type 3 Concentrate—for use undiluted from aircraft, ships or on beaches, using appropriate spray gear.

Notes: (1) Treatment rates for oil at sea would normally be one part dispersant to 2–3 parts of oil for Type 1 and Type 2 dispersant after dilution or one part of dispersant to 20–30 parts of oil for Type 3 dispersant applied neat.

(2) Use of a dispersant is conditional upon the prior approval of the local representative of the appropriate Government Department for any use in water of 20 m or less in depth, or within 1 mile of such depths; this includes use on beaches.

(3) All dispersants must be used in accordance with the manufacturers' instructions.

4 MATERIALS

(a) The oil spill dispersant shall consist of suitable ionic, non-ionic or a blend of such surfactants dissolved in a suitable solvent. It shall not contain compounds which could expose the user to an unacceptable toxicological hazard during normal spraying or handling operations when wearing a closely fitted face visor.

(b) It is unlikely that dispersant containing more than 3% wt of aromatics will pass the toxicity requirements. In addition the following ingredients are prohibited: benzene, chlorinated hydrocarbons, phenols, caustic alkali, and free mineral acid.

(c) The surfactants shall be wholly soluble in the solvent and shall remain distributed uniformly at all temperatures from $-10°C$ up to $50°C$ when stored for periods of up to 7 days.

SPECIFICATION FOR OIL SPILL DISPERSANTS—PERFORMANCE

(d) Type 2 dispersants shall be miscible with sea water at 1:10 concentration to form a solution or emulsion which has a viscosity not greater than that of the original dispersant alone.
Note: Type 1 and Type 3 dispersants have no requirement to be miscible with water since they will be applied undiluted.

5 FORMULATION

(a) Details of the formulation must be submitted (in confidence) to WSL and MAFF at the time of application for Qualification and Approval respectively. The details shall include the percentage, chemical name (when applicable), and function.
Note: For Qualification purposes WSL will send a form on which the information can be entered, when the contract for the work is sent.

(b) After a product has been qualified, no change in formulation will be permitted unless the change is approved in writing both by the Qualification Authority and by the Approval Authority.

6 QUALIFICATION TESTING

Note: The units quoted in this specification are consistent with the practice applied at present in the Chemical and Petroleum Industries.

(a) *Test Methods*
Unless otherwise stated, the test methods to be used shall be the latest published editions of those given in this Specification.

(b) *Tolerance of Test Methods*
Requirements contained herein are absolute and not subject to corrections for tolerance of test methods. If multiple determinations are made by the Qualification Authority, average results are to be used except for those test methods where repeatability data are given. In those cases, the average value derived from the individual results that agree within the repeatability limits given for the test method, may be used if the Authority permits.

(c) *Additional Test Requirements*
The Qualification Authority reserves the right to require additional testing of the product.

(d) A sample taken from any portion of the product shall comply with the requirements of Table A1.

TABLE A1

Test No.	Test	Type 1	Type 2	Type 3	Method
1	Appearance	Clear and homogeneous			Visual examination
2	Dynamic viscosity at 0°C, mPa s, maximum	50	250	250	Note 1
3	Flash point, °C, minimum	60	60	60	ASTM D93 IP 34 BS 2000 Part 34
4	Cloud point, °C (as received) maximum	Minus 10	Minus 10	Minus 10	ASTM D2500 IP 219
5	Efficiency index, % 2000 mPa s Fuel oil, minimum 500 mPa s Fuel oil, minimum	30 —	30 —	60 45	Annex to Appendix A
6	Storage Test (−10°C for 7 days)	Pass	Pass	Pass	Paragraph 4c
7	Miscibility with water	—	Pass	—	Paragraph 4d

Note 1: The viscometer supplied by UK Viscometer Ltd model no. UKLV8 is the preferred equipment. This is used in conjunction with the small sample adaptor and spindle No. TL-6.

GUIDELINES ON THE USE OF OIL SPILL DISPERSANTS

7 QUALITY ASSURANCE

(a) The manufacturer shall certify that each batch of the product is of the same formulation as that qualified in accordance with clauses 6 and 7 of this Specification.

(b) The Qualification Authority reserves the right to sample and test the product at any time. For bulk supplies, the Qualification Authority may require a 25 l reference sample to be taken at the time of blending.

(c) If any sample taken from a consignment is found not to comply with the requirements of this Specification, the whole consignment may be rejected.

(d) The provisions of this clause apply equally to the manufacturer and any sub-contractor.

8 KEEPING QUALITIES

The product, when suitably stored in its original sealed containers, shall retain the properties described in this Specification for a period, from the date of despatch, of not less than two years in temperate climates (-20 to $+30°C$).

9 CONTAINERS AND MARKING OF CONTAINERS

(a) The product shall be supplied in sound, clean, and dry containers, or bulk carriers suitable for the product and in accordance with the requirements of the contract or order.

(b) Coatings, paints, and markings of the containers shall comply with the requirements of the contract or order, and shall be to the satisfaction of the purchaser.

(c) It shall be the responsibility of the supplier to comply with any legal requirements for the marking of containers.

Warren Spring Laboratory
Stevenage
Herts
UK

April 1983, amended January 1986

ANNEX TO APPENDIX A

DETERMINATION OF EFFICIENCY INDEX OF OIL SPILL DISPERSANTS

1 SCOPE

This method estimates the quantity of a reference fuel oil that is dispersed into sea water by the application of an oil spill dispersant under standard laboratory test conditions. The method may also be used to estimate the quantities of crude oil and other petroleum products which could be dispersed into sea water.

2 SUMMARY OF METHOD

2.1 The oil spill dispersant is added dropwise to a measured volume of the test oil on the surface of sea water at 10°C contained in a conical separating funnel. The separating funnel is rotated about a horizontal axis at right angles to its longitudinal axis for a period of two minutes at 33 ± 1 rev/min. After rotation has ceased the flask is unstoppered and after one minute standing time 50 cm^3 of oily water are run off through the bottom tap. The quantity of oil in the water sample is then determined spectrophotometrically following extraction into chloroform. The test is carried out with two different viscosity grades of reference oil.

Note 1: The test is designed for determining the efficiency index for Type 3 dispersants to be applied undiluted. It may be adapted for measurements of the efficiency index for hydrocarbon-base and pre-diluted (1:10 water) dispersant concentrates (Types 1 and 2 dispersants) by increasing the quantity of dispersant added dropwise to the test oil to 2 cm^3.

3 DEFINITION

The Efficiency Index is defined as the percentage of the test oil which has been transferred as small droplets into the water phase under the conditions of the test, assuming complete and even distribution at the time of sampling.

4 APPARATUS

4.1 *Separating funnels*

4.1.1 A conical separating funnel to BS 2021, with a nominal capacity of 250 cm^3 and complying with the dimensions shown in Fig. A.1.

4.1.2 A separating funnel to BS 2021, with a nominal capacity of 100 cm^3.

Note 2: Before use funnels should be thoroughly cleaned with hot water and a laboratory detergent. When all traces of organic matter have been removed, and the glass is water wetted, the flask should be rinsed thoroughly with hot water and then with distilled water before being left to drain.

4.2 *Motor driven rack*—A motor driven rack into which the 250 cm^3 separating funnel can be clamped and rotated at 33 ± 1 rev/min about a horizontal axis approximately 80 mm below the top of the funnel. A photograph of suitable equipment is shown in Fig. A2.

Note 3: A significant alteration of the position of the horizontal axis and any movement of the flask other than smoothly about a horizontal axis may affect the results.

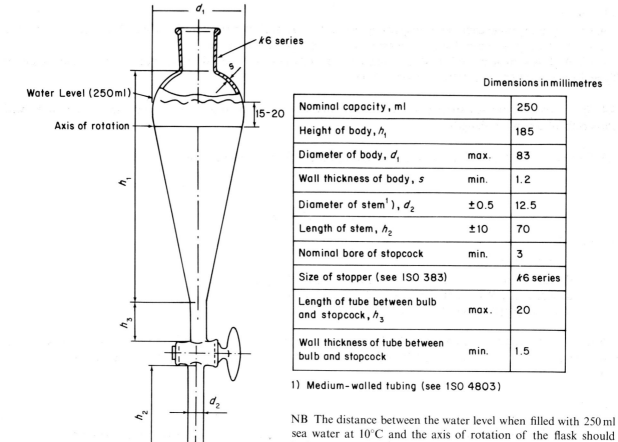

Dimensions in millimetres		
Nominal capacity, ml		250
Height of body, h_1		185
Diameter of body, d_1	max.	83
Wall thickness of body, s	min.	1.2
Diameter of stem[1]), d_2	±0.5	12.5
Length of stem, h_2	±10	70
Nominal bore of stopcock	min.	3
Size of stopper (see ISO 383)		*k*6 series
Length of tube between bulb and stopcock, h_3	max.	20
Wall thickness of tube between bulb and stopcock	min.	1.5

[1]) Medium-walled tubing (see ISO 4803)

NB The distance between the water level when filled with 250 ml sea water at 10°C and the axis of rotation of the flask should be between 15 and 20 mm.

Figure A1. Standard conical separating funnel 250 ml capacity.

Figure A2. Motor driven rack with separating funnel. Biggs Wall Fabricators Ltd, Hampden House, Arlesey, Beds SG15 6RT.

SPECIFICATION FOR OIL SPILL DISPERSANTS—PERFORMANCE

4.3 *Syringes*

4.3.1 A glass syringe capable of dispensing accurately $0.2\,cm^3$ of dispersant in drops of between 5 and $10\,\mu l$.

4.3.2 A glass syringe capable of dispensing accurately $5.0\,cm^3$ of the test oil.

4.3.3 A glass syringe capable of dispensing accurately $2.0\,cm^3$ of Type 1 dispersant or $2.0\,cm^3$ of the prediluted Type 2 dispersant (diluted 1:10 with sea water).

4.4 *Spectrophotometer*—A spectrophotometer capable of measuring absorbance at 580 nm and equipped with glass cells of 5 mm pathlength.

4.5 *Stopclock*—A stopclock capable of measuring up to 10 min in 1 s intervals.

5 REAGENTS

5.1 *Test oils*

5.1.1 A reference fuel oil which has the following characteristics:

dynamic viscosity $10°C$: $1800-2200\,mPa\,s$ at $10\,s^{-1}$ shear
asphaltenes (IP 143/84): 1.5% by weight maximum
pour point (IP 15/67): below $5°C$.

5.1.2 A lower viscosity reference fuel oil made by diluting the test oil at para 5.1 with kerosine (para. 5.2) to provide an oil with a viscosity at $10°C$ of $500\,mPa\,s$ at $10\,s^{-1}$ shear rate. Usually 2–5% addition of kerosine is sufficient.

Note 4:
(i)   A blend of fuel and diesel oil is permissible.
(ii) The dynamic viscosity can be measured conveniently using a Ferranti Portable Viscometer Type VM.
(iii) A suitable test oil may be purchased from the Warren Spring Laboratory[1] in small quantities.

5.2 *Kerosine*—Odourless kerosine to BS 2869 Class C1.

5.3 *Sea Water*—Water with a salinity between 33 and 35 g/kg. Either sea water taken from the sea or synthetic sea water[2] is permissible.

5.4 *Chloroform*—AR grade. (CAUTION: *Chloroform is a toxic, volatile chemical which is absorbed by inhaling the vapour or through the skin by contact with the liquid. In addition, when heated to decomposition, it breaks down to the highly toxic phosgene gas. Handle in a fume cupboard or similar well-ventilated area. Do not smoke.*)

5.1 *Sodium sulphate, anhydrous*—Reagent grade.

6 METHOD

6.1 *Calibration*—Transfer 0.2, 0.5, 0.7 and 0.9 g of the test oil, accurately weighed, into separate $100\,cm^3$ volumetric flasks. Add $50\,cm^3$ of chloroform to each flask and mix well to dissolve the oil. Make up to the mark with chloroform, stopper and mix well.

Measure the absorbance of each solution at 580 nm in glass cells at 5 mm path-length. Prepare a graph of absorbance against the weight of oil contained in $100\,cm^3$ of the chloroform solution. Repeat the calibration with the second reference fuel oil.

Note 5: The graphs should each be a straight line.

6.2 *Procedure*—Add to the $250\,cm^3$ separating funnel $250\,cm^3$ of sea water cooled to $10\pm0.5°C$. This temperature should be maintained throughout the test by conducting the work in a suitably temperature controlled chamber.

---
[1] Warren Spring Laboratory, Department of Industry, PO Box 20, Gunnels Wood Rd, Stevenage, Hertfordshire SG1 2BX.
[2] A suitable synthetic sea water is described in IP 135.

Place the separating funnel in the rotatable rack. Leave unstoppered. Transfer 5 cm^3 of the test oil to the water surface, weighing the syringe before and after use to calculate the weight of 5.0 cm^3 of the oil. Start the stopclock and allow the transferred oil to rest on the surface of the water for one minute. Using the appropriate syringe transfer the requisite amount of dispersant (Note 6) to the surface of the oil in the separating funnel so that the drops are distributed as evenly as possible over the surface of the oil (Note 7). Replace the stopper.

When the time on the stopclock shows 2.5 min from the addition of the oil, start the rotation of the separating funnel and continue at 33 ± 1 rev/min for a further 2 min.

Stop the separating funnel in the upright vertical position and allow the contents to stand undisturbed for exactly 1 min before running off 50 cm^3 of the oily water through the tap and collecting it in a 50 cm^3 measuring cylinder. The time taken for this operation should not exceed 10 s.

Transfer the oily water from the measuring cylinder to the 100 cm^3 separating funnel. Wash the measuring cylinder twice with 10 cm^3 of chloroform and add the washings to the separating funnel. Stopper the funnel and shake for 1 min. Allow the phases to separate completely and run off the chloroform layer through a No. 1 Whatman filter paper containing anhydrous sodium sulphate. Repeat the chloroform extraction twice more using 20 cm^3 of chloroform on each occasion. Wash the filter with 20 cm^3 of chloroform and combine the dried extracts and washings in a 100 cm^3 volumetric flask. Make up to the mark, stopper and mix well.

Measure the absorbance of the chloroform extract against a chloroform blank at 580 nm in glass cells of 5 mm path length. Using the calibration graph calculate the weight of oil contained in the 50 cm^3 oily water sample. Repeat the measurements to obtain three separate determinations on each of the two reference fuel oils.

Note 6: For a Type 1 dispersant 2.0 cm^3 of dispersant are added from a syringe.

For a Type 2 dispersant 5.0 cm^3 of concentrated dispersant should be thoroughly mixed with 45.0 cm^3 of sea water. For the test 2.0 cm^3 of the prediluted dispersant are added from a syringe.

For a Type 3 dispersant 0.20 cm^3 of undiluted dispersant are added from a micrometer syringe.

Note 7: The dispersant should be added dropwise starting at the centre of the layer of oil and radiating outwards to give an even distribution over the surface of the lens. Should the dispersant touch the air/water interface and cause the oil to migrate to the side of the separating funnel before completion of the dispersant addition, reject the test and start again. When 2 cm^3 of a type 1 or 2 dispersant is added dropwise onto the oil there is usually some run-off into the water which has to be accepted.

7 CALCULATION

7.1 The Efficiency Index, E, is calculated from the following equation:

$$E = \frac{\text{weight of oil in 50 cm}^3 \text{ sample of oily water} \times 500}{\text{total weight of oil added to 250 cm}^3 \text{ separating funnel}} \%$$

8 REPORT

8.1 The Efficiency Index which is reported is the average of three separate determinations. The calculated average is reported to the nearest whole percentage for each of the two reference fuel oils. (IP 83-009).

APPENDIX B

SPECIFICATION FOR OIL SPILL DISPERSANTS—TOXICITY

In order to be included in the list of approved products, a dispersant must pass toxicity tests, which are carried out at the Ministry of Agriculture, Fisheries and Food's (MAFF) Fisheries Laboratory at Burnham-on-Crouch. Two laboratory tests have been developed to reflect the different conditions of use of dispersants at sea and on rocky shores. The 'sea' test reflects the conditions which may occur when a dispersant is applied to oil floating on the surface of the water; a WSL Qualified product which passes this test will be approved for use at sea and also on sandy/pebble amenity beaches. If a dispersant is to be used on rocky shores, it must also pass the 'beach' test to gain approval.

1 'SEA' TEST

1.1 Rationale

When dispersants are properly applied to oil spilt at sea, marine organisms are exposed, not to dispersant alone, but to a mixture of oil and dispersant. Wind and wave action may also result in some untreated oil being physically dispersed into the water column. The 'sea' test is based on a comparison of the toxicity of an oil/dispersant mixture with that of a physical dispersion of oil; these tests are made simultaneously under identical conditions. A comparative test of this type has the advantage of eliminating variation between the results of tests carried out on different occasions, which could be caused by factors such as seasonal or other changes in sensitivity of the test organisms. The toxicity is measured as the mortality of brown shrimps (*Crangon crangon* L), following 100 min exposure to $1000 \mu l\, l^{-1}$ of oil alone, or to a similar amount of oil with the dispersant under test. Fresh Kuwait crude oil is used as the reference substance because it is a typical Middle Eastern crude oil and its effects on marine life are now well documented. The high oil concentration is necessary to produce a measurable toxic effect in a short time; a lower concentration would require a longer exposure period, during which the toxicity of the oil would be expected to change through 'weathering'. Mortalities are not counted immediately but at the end of a 24 h 'recovery' period in clean flowing sea water to avoid animals which have been temporarily anaesthetised by the oil being counted as dead. A dispersant passes the 'sea' test if it does not produce a dispersant/oil mixture which is significantly more toxic than the oil alone (see section 3 of this appendix).

1.2 Materials

The sea water used for the tests and for maintaining the test animals during acclimation is taken from the estuary of the River Crouch on a flood tide. It is stored in settling tanks for about two days then pumped to the laboratory's header tank where it is brought to the test temperature ($15°C \pm 1°C$) before being supplied to the test animal stock tanks. A $10 \mu m$ in-line membrane filter removes any remaining silt particles from the water which is used to fill the test tanks. The salinity of the water, which is measured daily, ranges from 28 to 35 parts per thousand (‰).

Sub-samples of the reference fresh Kuwait crude oil are stored at ambient temperature in 250 ml air-tight metal cans to prevent losses due to evaporation. A fresh sub-sample is used for each experiment and the oil is brought to the test temperature before use.

GUIDELINES ON THE USE OF OIL SPILL DISPERSANTS

Samples of dispersant for toxicity testing under the Food and Environment Protection Act are also obtained from the Warren Spring Laboratory and are always taken from the same batch as that used for the efficiency tests. The dispersant is stored at ambient temperature and brought to the test temperature before use. Concentrate dispersants are diluted immediately before use with sea water to give a 10% (v/v) solution.

1.3 Experimental animals

Adult brown shrimps (*Crangon crangon* L) are caught in the estuary of the River Crouch. On arrival in the laboratory, the animals are transferred to shallow 40 l polyethylene stock tanks at a maximum density of 200 shrimps per tank and any dead or injured animals are removed. The animals are maintained in well aerated, gently flowing sea water for 2–4 days before the start of the test. They are not fed during their period in the laboratory, which does not exceed 6 days. Healthy shrimps of between 50 and 70 mm total length, excluding antennae (weight about 1–3 g), are selected for the toxicity tests; freshly moulted shrimps are excluded.

1.4 Apparatus

A 'sea' test tank is illustrated in Figure B1. This consists of a cylindrical 'Perspex' tank in which 18 l of sea water are stirred by means of a three-bladed propeller enclosed in a central 'Perspex' column. The propeller is driven, via a magnetic coupling, by a pneumatic motor, the speed of which is adjusted (with the aid of a tachometer) to between 1350 and 1450 rev/min. This draws the oil (or oil plus dispersant) into the column through an aperture level with the water surface and expels it through another aperture close to the bottom of the tank thus producing a uniform dispersion of small (<1 mm) oil droplets without drawing air into the system or causing undue physical stress to the test organisms. It also maintains the dissolved oxygen concentration at close to air saturation value (about 8.4 mg l^{-1}) so that no extra aeration is required.

The rectangular 'Perspex' recovery tanks, to which the animals are transferred, hold 10 l of sea water flowing at about 10 l h^{-1}. These are aerated by means of a dropper pipette connected to the laboratory's compressed air supply.

The room in which the tanks are housed is maintained at a constant temperature of 15°C ($\pm 1°C$) and has a 12 h light: 12 h dark photoperiod.

1.5 Procedure

Ten test tanks are filled with 18 l of sea water and aerated for at least 1 h. Twenty shrimps are then randomly added to each of the tanks and the lids put in place. After a further 2 h the aerators and lids are removed from

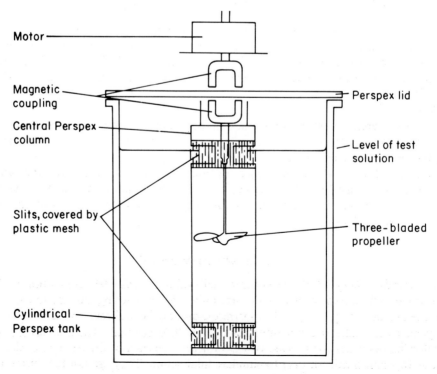

Figure B1. 'Sea' test agitation tank.

the tanks to allow the test material to be added. In each tank, 18 ml of oil are poured directly onto the surface of the water from a measuring cylinder to give a nominal oil concentration of $1000\,\mu l\,l^{-1}$. This is followed in five of the tanks by 18 ml of dispersant (neat conventional or 10% dilution of concentrate) distributed as evenly as possible over the oil surface, thus giving dispersant:oil application rates of 1:1 for conventional (hydrocarbon solvent based) dispersants and 1:10 for concentrates. One minute after adding the dispersant the lids are replaced to cover the tanks and the stirrer motors are switched on.

After 100 min the motors are switched off, the lids removed, and the oily water siphoned out of the tanks. As soon as each tank is emptied, the test animals are gently transferred to 10 l tanks of clean, gently flowing, aerated sea water for 24 h so that those which are lightly anaesthetised by the oil can recover. At the end of this period, the number in each tank found to be dead (defined as lack of response to gentle prodding) is recorded. Because cannibalism of freshly moulted shrimps can occur, the number of animals in each tank remaining alive at the end of 24 h is also noted so that the mortalities caused by the test treatment can be calculated. The mortalities in the dispersant treated tanks (test) are then compared statistically with those in the oil (reference) to determine whether the addition of the dispersant had a significant effect on the toxicity of the oil.

2 'BEACH' TEST

2.1 Rationale

If oil is stranded on rocky shores and a decision is taken to remove it by chemical means, it is possible that dispersants will be sprayed directly on to intertidal organisms such as winkles and limpets. For this reason the effect measured in the 'beach' test (now more properly called the 'rocky beach' test) is mortality of a typical intertidal organism, the common limpet (*Patella vulgata* L). The test is based on a comparison of the toxicity of dispersant with that of oil alone, the tests being carried out at the same time and under identical conditions. The limpets are sprayed with oil (reference) or dispersant (test) at $0.4\,l\,m^{-2}$ (the average recommended dispersant application rate), left in air for 6 h (the average exposure likely to occur when sprays are applied randomly throughout the tidal cycle), then rinsed with clean sea water and placed in recovery tanks with a simulated tidal cycle. As in the 'sea' test the reference substance used is fresh Kuwait crude oil.

A dispersant passes the 'beach' test if the dispersant (test) is not significantly more toxic than the oil (reference). Since both dispersant and oil are likely to be washed off by the incoming tide, where they could affect littoral communities, dispersants are only approved for rocky shore use if they also pass the 'sea' test.

2.2 Materials

The sea water used for the tests and for maintaining the test animals during acclimation is taken from the estuary of the River Crouch on a flood tide. It is stored in settling tanks for about two days then pumped to the laboratory's header tank where it is brought to the test temperature ($15°C \pm 1°C$) before being supplied to the test animal stock tanks. A 10 μm in-line membrane filter removes any remaining silt particles from the water which is used to fill the test tanks. The salinity of the water, which is measured daily, ranges from 28 to 35 parts per thousand (‰).

Sub-samples of the reference fresh Kuwait crude oil are stored at ambient temperatures in 250 ml air-tight metal cans to prevent losses due to evaporation. A fresh subsample is used for each experiment and the oil is brought to the test temperature before use.

Samples of dispersant for toxicity testing under the Food and Environment Protection Act are always taken from the same batch as that used for the efficiency tests. The dispersant is stored at ambient temperature and brought to the test temperature before use. Concentrate dispersants are diluted immediately before use with sea water to give a 10% (v/v) solution.

2.3 Experimental Animals

Common limpets (*Patella vulgata* L) of 30 to 40 mm shell width, are collected from a beach with chalk boulders where they can be carefully prised off with an oyster knife without causing damage to the shells. On return to the laboratory, the animals are placed, shell uppermost, on polythene sheets (to facilitate subsequent removal) in 40 l polyethylene stock tanks at a maximum density of 1000 per tank. The animals are maintained in well aerated, gently flowing sea water for at least two days before the start of the test and subjected to intermittent immersion (about 18 h immersed: 6 h dry) to simulate tidal action. They are not fed during their period in the laboratory, which does not exceed two weeks.

2.4 Apparatus

For the purpose of the test, the limpets are placed on, and become attached to, rectangular 440 cm² test plates made of 6 mm 'Perspex' with sharply bevelled edges to deter the animals from moving from one side to another. Each plate has two stainless steel hooks by which it can be suspended vertically in a recovery tank.

Test materials are sprayed over the plate area from a hand operated sprayer. During spraying, and for 6 hours afterwards, the plates are suspended horizontally across parallel rods in tanks which are large enough to hold five plates side by side, with enough space between each to prevent the animals from transferring from one to another. Each tank is covered by a lid between the time when the plates are sprayed and their subsequent transfer to the recovery tank.

The rectangular 'Perspex' recovery tanks hold 10 l of sea water flowing at about 10 l h^{-1}. These are subjected to intermittent immersion as described in section 2.3 of this appendix. Aeration is by means of a dropper pipette connected to the laboratory's compressed air supply.

The room in which the tanks are housed is maintained at a constant temperature of 15°C ($\pm$1°C) and has a 12 h light:12 h dark photoperiod.

2.5 Procedure

The day before the start of the test, 10 test plates are suspended horizontally on racks in the stock tanks. About 24 limpets are then placed on each plate and left overnight in well aerated, gently flowing sea water to attach themselves. Just before the start of the test, the number of animals per plate is reduced to 20, selecting for removal any animals with chipped or excessively curved bases to their shells. The plates are then placed horizontally, attached animals uppermost, in two spraying tanks. Each limpet on each plate in one of the tanks is sprayed, from a height of about 10 cm, with 0·8 ml of oil to give an application rate of 0·4 l m^{-2}. The second set of plates is sprayed in a similar manner with dispersant. A small amount of sea water is added to each tank, which is then covered with a lid to maintain the humidity.

After 6 h, each plate of sprayed limpets is removed from the spraying tank, washed for 15 sec with clean sea water and suspended vertically in a recovery tank. Further rinsing takes place in the recovery tanks during periods in which the plates are immersed. Limpets which become detached from the plates are recorded as dead and removed from the tanks 1, 24 and 48 h after rinsing. After 48 h, any remaining limpets not firmly attached to their plates are gently detached and placed on the tank floor. Those failing to re-attach to a surface within a further 24 h period are also counted as dead. The total mortalities after 72 h on the five plates treated with dispersant (test) are then compared statistically with those on the plates treated with oil (reference) to determine whether the dispersant was significantly more toxic than the oil.

3 STATISTICAL ANALYSIS OF TOXICITY TEST DATA

The 'sea' and 'beach' test data are analysed using a programme written for an Apple IIe computer. The homogeneity of the two sets of replicates is tested to determine whether or not the animals in different tanks with the same treatment had the same probability of mortality. If either set of replicates (test or reference) fails the homogeneity test, the toxicity test has to be repeated. If the two data sets are homogeneous, the significance of any difference between test and reference mortalities is determined using an unpaired comparison of two proportions (e.g. Armitage, 1971). A dispersant fails an experiment if the dispersant plus oil mixture ('sea' test) or dispersant alone ('beach' test) is shown to be significantly more toxic than the reference oil.

3.1 Homogeneity Test

The homogeneity of each set of n (=5) replicates is tested by constructing a 2× contingency table, as follows:

$$O \text{ (dead)} = \text{observed number dead in each tank}$$
$$O \text{ (alive)} = \text{observed number alive in each tank}$$
$$N = O \text{ (dead)} + O \text{ (alive)}$$
$$E \text{ (dead)} = \text{expected number dead in each tank}$$
$$= \frac{\sum O \text{ (dead)}}{\sum N} \times N$$

SPECIFICATION FOR OIL SPILL DISPERSANTS—TOXICITY

E (alive) = expected number alive in each tank

$$= \frac{\sum O(\text{alive})}{\sum N} \times N$$

$$\chi^2 = \frac{\sum [O(\text{dead}) - E(\text{dead})]^2}{E(\text{dead})} + \frac{\sum [O(\text{alive}) - E(\text{alive})]^2}{E(\text{alive})}$$

with n-1 (=4) degrees of freedom.

The replicates are considered to be homogeneous if χ^2 is not significant at the 95% probability level.

3.2 Significance Test

The significance of any difference between test and reference mortalities is tested as follows:

	Reference	Test
Total number dead =	r_1	r_2
Total number tested =	n_1	n_2
Proportion dead	$p_1 = \frac{r_1}{n_1}$	$p_2 = \frac{r_2}{n_2}$

Pooled proportion, $p = \frac{r_1 + r_2}{n_1 + n_2}$

Difference between mortalities, $d = p_1 - p_2$

95% confidence limits of $d = d \pm 1{\cdot}96 \sqrt{\left[\frac{p_1(1-p_1)}{n_1} + \frac{p_2(1-p_2)}{n_2}\right]}$

Standardised normal deviate, $u = \dfrac{p_1 - p_2}{\sqrt{\left[p(1-p)\left(\dfrac{1}{n_1} + \dfrac{1}{n_2}\right)\right]}}$

the mortalities are considered to be significantly different if u is significant at the 95% probability level.

4 REFERENCES

Armitage, P., 1971. Statistical methods in medical research. Blackwell Scientific Publications, Oxford and Edinburgh; 500 pp.

Blackman, R. A. A., Franklin, F. L., Norton, M. G. and Wilson, K. W., 1977. New procedures for the toxicity testing of oil slick dispersants. Fish. Res. Tech. Rep., MAFF Direct. Fish. Res., Lowestoft, (39); 7 pp.

APPENDIX C

APPLICATION EQUIPMENT FOR DISPERSANT CONCENTRATES

Since 1983 when the revised dispersant specification was issued (Appendix A), the use of dispersants has been mainly directed towards the application of undiluted Type 3 products using equipment specially designed for these materials. Consequently, this Appendix describes generally, under the heading of Aerial, Sea and Beach Applications, the types of equipment which have been found suitable. The treatment rates have been described in Section 2.

1 AERIAL APPLICATION

1.1 Fixed Wing Aircraft

Purpose-built aircraft designed for agricultural spraying operations can, if suitably modified, be used for dispersant application and have the advantage of being capable of using short or improvised landing strips. Their size and manoeuvrability make them ideally suited for dealing with small spills or fragmented slicks close to shore. However, they are usually single-engined aircraft with small payloads and are restricted by the distance they can operate safely from the coast. Larger multi-engined aircraft offer the required range, payload, speed and safety for the treatment of large spills offshore. A number of different types of aircraft have been converted for spraying and some of these are listed below:

AIRCRAFT TYPE		DISPERSANT CARRYING CAPACITY (LITRES)	MAX TRANSIT SPEED (KNOTS)	MIN RUNWAY LENGTH (METRES)
Beach Baron	2 piston	450	200	410
BN Islander	2 piston	1000	140	170
BN Trislander	3 piston	1250	145	395
Canadair CL 215	2 piston	5300	160	915
DC3	2 piston	4600	130	1000
DC4	4 piston	9460	190	1525
DC6	4 piston	13250	210	1525
Piper Aztec	2 piston	570	175	300
Short Sky Van	2 turbine	1200	170	510
Twin Otter	2 turbine	2100	170	320
Volpar Turbo Beech 18	2 turbine	1100	220	510

Illustrations showing systems which have been developed for use in the Britten Norman Islander and DHC-6 Twin Otter are shown in Figs C1 and C2.

APPLICATION EQUIPMENT FOR DISPERSANT CONCENTRATES

Figure C1. Britten-Norman Islander. Harvest Air Ltd, Municipal Airport, Southend on Sea, Essex SS2 6NN.

Figure C2. Spray boom fitted to DHC Twin Otter.

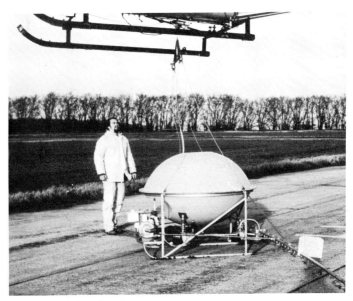

Figure C3. TC-3 helicopter dispersant spray bucket. Rotortech Ltd, Bourn Airfield, Cambridge CB3 7TQ.

GUIDELINES ON THE USE OF OIL SPILL DISPERSANTS

Where the aircraft spray system is not specifically designed for oil spill dispersants, calibration checks should be carried out before use (see Sections 2.1.2 amd 3.1.2).

There are, in addition, oil dispersant spray systems which can be used in conjunction with non-dedicated aircraft.

1.2 Helicopters

Most helicopters are able to carry underslung loads and can therefore be equipped with a 'bucket' type spray system such as depicted in Fig. C3 without the need for modification.

The larger twin engined helicopters offer the range, payload, speed and safety for operation at sea. However, single engined helicopters can be used provided they are equipped with suitable flotation devices.

Some helicopters thought suitable for dispersant spraying are listed below.

HELICOPTER TYPE	EXTERNAL LOAD CAPACITY kg	CRUISING SPEED KNOTS	EST. MAX. OFFSHORE RADIUS NAUTICAL MILES
Aerospatiale Super Puma 332	4428	120	175
355 E	886	100	100
355 F	886	100	175
365 C	1360	125	116
365 N (Dauphin 2)	1600	135	190
Bolkow B105	670	110	155
Bell 212	2232	125	110
Sikorsky S61	4464	112	156
S76 (Mk 2)	1786	125	200
Westland 30 Series	2008	120	190

2 SEA APPLICATION

Various manufacturers have developed equipment suitable for spraying undiluted Type 3 dispersant from vessels. Examples of equipment are shown in Figs C4–C6. When using these systems in calm conditions (i.e. wind speeds less than 10 knots), consideration should be given to providing agitation to the sea by towing breaker boards or similar devices.

Figure C4. Biggs Wall wide spray boom for Type 3 concentrates applied undiluted. Biggs Wall Fabricators Ltd, Hampden House, Arlesey, Beds SG15 6RT.

APPLICATION EQUIPMENT FOR DISPERSANT CONCENTRATES

Figure C5. Frank Ayles Clearspray HP-2 for Type 3 dispersants applied undiluted. Frank Ayles & Associates, 120 Whitechapel High St, London E1 7PT.

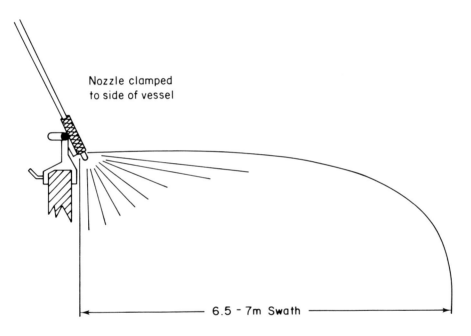

Figure C6-A. Cooper Pegler 'AR100D Super' for Type 3 dispersants applied undiluted. ALBA OPD Ltd, Unit 58, Sherwood Rd, Aftonfield, Bromsgrove, Worcs, B60 5DR.

Figure C6-B. General view of Cooper Pegler 'AR 100D Super'.

GUIDELINES ON THE USE OF OIL SPILL DISPERSANTS

3 BEACH APPLICATION

The most appropriate equipment for this application depends on the type of beach material to be cleaned, the ease of access and the scale of the operation. For small inaccessible beaches, portable back-pack sprayers are the most suitable. For large expanses of beach, purpose-built vehicles, tractors or aircraft can be used. Examples of different equipment are shown in Figs C7–C9.

Figure C7. Hand operated dispersant bandpack. Biggs Wall Fabricators Ltd (see Fig. C4) or ALBA OPD Ltd (see Fig. C6-A).

Figure C8. Invictacat multi-terrain dispersant spray vehicle. The Crayford Special Equipment Company, High Street, Westerham, Kent TN16 1RG.

APPLICATION EQUIPMENT FOR DISPERSANT CONCENTRATES

Figure C9. Cooper Pegler Beachguard Super Oil Dispersant Trailer (for Type 1 dispersants). ALBA OPD Ltd, Unit 58, Sherwood Rd, Aftonfield, Bromsgrove, Worcs B60 5DR.

4 MAN-MADE STRUCTURES

Oil deposited on the sides of structures such as piers, walls, jetties and promenades can be removed by the application of a gelled dispersant followed by high pressure sea water washing to disperse the oil and wash it into the sea. The gel is normally left in contact with the oil for as long as possible (6–24 hours). The dispersant is gelled by bringing a solution of a chemical into contact with the dispersant as it leaves the nozzle attached to a mixing chamber. Suitable equipment is shown in Fig. C10. Details of the technique are given in the Warren Spring Laboratory report LR 216.

Figure C10. Dispersant gel equipment. Biggs Wall Fabricators Ltd, Hampden House, Arlesey, Beds SG15 6RT.

APPENDIX D

AUTHORITIES APPROVING THE USE OF DISPERSANTS IN UK WATERS

England and Wales	Ministry of Agriculture, Fisheries and Food Fisheries Division 1C Marine Pollution Branch Great Westminster House Horseferry Road London SW1P 2AE Telephone: 01 216 6311 Telex 21271
Scotland	Department of Agriculture and Fisheries for Scotland Chesser House Gorgie Road Edinburgh EH11 3AW Telephone: 031 443 4020 Telex 72162
Northern Ireland	Department of the Environment for Northern Ireland Parliament Building Stormont Belfast BT4 3SS Telephone: Belfast (0232) 63210 Telex 748025

APPENDIX E

BIBLIOGRAPHY

SUGGESTIONS FOR FURTHER READING

1. IMO/UNEP Guidelines on Oil Spill Dispersant Application and Environmental Considerations. IMO/UNEP. 1982.
2. Technical Information Paper No. 4—Use of Oil Spill Dispersants. ITOPF 1982.
3. Technical Information Paper No. 3—Aerial Application of Oil Spill Dispersants. ITOPF. 1982.
4. A Field Guide to Coastal Oil Spill Control and Clean-Up Techniques, Report No. 9/81. CONCAWE. September 1981.
5. Cormack, D. Response to Oil and Chemical Marine Pollution. Applied Science Publishers, 1983. ISBN 0-85334-182-6.
6. Morris, P. R. and Martinelli, F. N. A Specification for Oil Spill Dispersants. Warren Spring Laboratory Report No. LR 448 (OP). May 1983. Also published by the Institute of Petroleum (U.K.) as publication IP 83-009 under the same title.
7. Crowley, S. and Nightingale, J. Evaluation of Oil Spill Dispersant Concentrates for Beach Cleaning. Warren Spring Laboratory Report No. LR 463 (OP). November 1983.
8. Crowley, S. Shipboard Spraying Equipment for Undiluted Dispersant Concentrates. Warren Spring Laboratory Report No. LR 492 (OP). July 1984.
9. Cormack, D. The Use of Aircraft for Dispersant Treatment of Oil Slicks at Sea. Report of a Joint UK Government/Esso Petroleum Investigation. September 1983.
10. Martinelli, F. N. Dispersant Spraying Equipment: the Rotortech TC3 Underslung Spray Bucket. Warren Spring Laboratory Report LR 407 (OP). November 1981.
11. Croquette, J. The Use of Dispersants at Sea to Control Oil Slicks. CEDRE Publication No. R.85.70.E. May 1985.
12. Oil Spill Clean-up of the Coastline—A Technical Manual. Warren Spring Laboratory, on behalf of the Department of the Environment, UK. 1982. ISBN 0-85624-262-4.
13. Oil Spill Response—Options for Minimising Adverse Ecological Impacts. American Petroleum Institute, Publication No. 4398. August 1985.